14.98

INTERIM SITE

M

Our World

WEATHER AND CLIMATE

John Mason

Silver Burdett Press
Englewood Cliffs, New Jersey

Titles in this series

First published in 1988 by
Wayland (Publishers) Ltd
61 Western Road. Hove
East Sussex BN3 1JD

Adapted and first published
in the United States in 1991 by
Silver Burdett Press,
Prentice Hall Building,
Englewood Cliffs, New Jersey 07632

Consulting Editors : Lee Ann Srogi
and Professor Hickox

ISBN 0-382-24225-4

Typeset by DP Press, Sevenoaks, Kent
Printed in Italy by G. Canale & C. S.p.A., Turin

Front cover, main picture A frosty winter's day.

Front cover, inset Lilies in the rain.

Back cover Lightning flashes illuminate the night sky during a violent thunderstorm.

Contents

Weather and climate on earth

Everyone is aware of the weather. It gives us sunny days, showers of rain, blizzards, and thunderstorms. The weather affects our lives and as a result is of great interest to people everywhere. In areas where the weather is changeable and unpredictable, one of the main topics of conversation is what the weather will do next.

In some parts of the world the weather is a matter of life and death. Consider the effects of the violent winds of the typhoons of the China Sea, the hurricanes of the Caribbean or the tornadoes of the midwest states of North America. Such events can have serious consequences for thousands of people unfortunate enough to get caught up in their wake. For those living in places like Africa or India, where the weather is generally less varied and more predictable, the failure of the eagerly awaited rainstorms of the summer monsoon can lead to drought and loss of the year's crops, triggering widespread famine. In some cases it takes years to recover from the consequences.

It is usually bad weather that makes headlines – a major storm devastating some island in the tropics, a particularly cold winter in North America, a famine-inducing drought in Ethiopia, or a heatwave leading to fierce fire-storms in southern Australia. We can see that the weather is very varied all over the world. In some places the weather is nearly always the same, such as in those deserts where it is hot and dry all year round. In other places, such as Britain, the weather can be different every day of the year.

The average weather conditions experienced by a place over a long period of time are known as the climate. Is the earth's climate getting steadily warmer, or are we heading for another ice age where the polar ice sheets will dramatically increase in size? Trying to work out which of these two is the more likely is not as easy as you might think, because at any time the weather may be rather different from the "normal" given by the climatic description.

In this view of earth from space, the oceans and continents are partly hidden by swirling masses of cloud.

The ever-changing weather affects the daily lives of people all over the world.

Above These floods in Brazil have been caused by heavy tropical rains. Flooding causes an enormous amount of damage and hardship each year.

Left A gasoline tanker brings much-needed water to a drought-stricken area in Ethiopia.

The seasons

Throughout the year, the earth moves around the sun along a path called its orbit, which is not quite circular. This means that at certain times of the year the earth is slightly closer to the sun than at others. As it rotates, the earth spins on its axis. The axis is not upright but leans over at an angle of 23.5 degrees to the vertical. It is this tilt that causes the seasons of spring, summer, autumn and winter that we experience on the earth. If one hemisphere is tilted toward the sun, the other hemisphere is tilted away from the sun.

When the northern hemisphere is tilted away from the sun, it is winter in the north. The sun is then lower in the sky than it is in summer, so it spends less time above the horizon and the days are shorter. The sun's rays also strike the ground at a shallow angle and are spread over a wider area, so their heating effect is less. When the sun is fairly low in the sky, its rays have to pass through a greater thickness of the atmosphere, which further weakens them. The weakness of the sun's rays, combined with the shorter days, makes winter colder than summer.

While the northern hemisphere freezes in winter, the southern hemisphere enjoys warmer temperatures. However, the northern and southern polar zones do not have a real summer. While the summer is warmer than winter, it is never hot in the polar zones. Although there are long hours of daylight, the sun is always weak and its rays are never strong enough to melt the polar ice caps. Near the equator, however, the sun is high in the sky throughout the year; therefore, winter is only a little cooler than summer. The greatest seasonal difference in many areas near the equator are dry periods and rainy periods.

Between the tropics and the poles are the temperate regions, such as Europe and North America. These places experience both hot and cold weather, and most have four distinct seasons. These seasons gradually change from winter to spring as the days get longer and warmer. Spring is followed by summer. As the days get shorter it gradually gets colder, and summer gives way to autumn. Winter returns during the period of the shortest days.

The seasons vary greatly in the temperate regions, as these summer and winter scenes show.

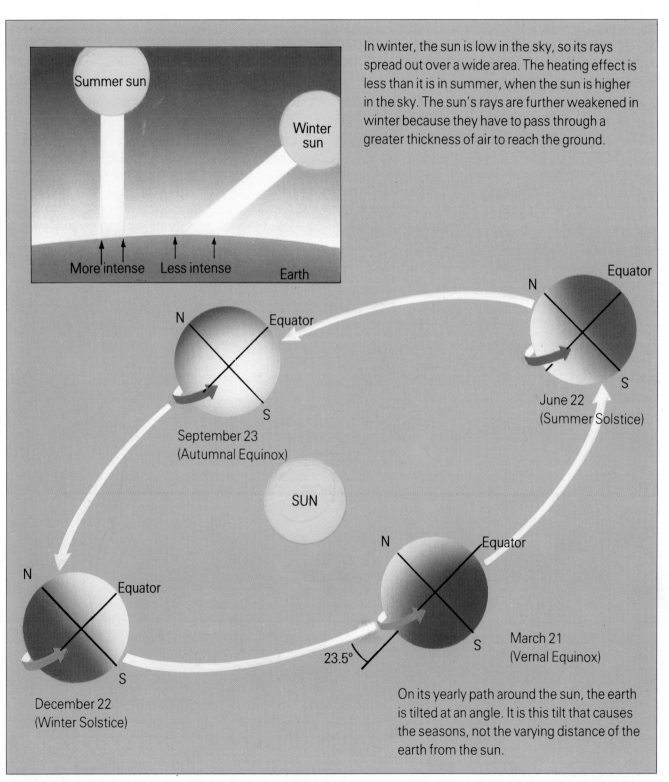

In winter, the sun is low in the sky, so its rays spread out over a wide area. The heating effect is less than it is in summer, when the sun is higher in the sky. The sun's rays are further weakened in winter because they have to pass through a greater thickness of air to reach the ground.

Summer sun

Winter sun

More intense   Less intense   Earth

Equator
N
June 22
(Summer Solstice)

N   Equator
September 23
(Autumnal Equinox)
S

SUN

N   Equator
S
March 21
(Vernal Equinox)

N
Equator
S
December 22
(Winter Solstice)

23.5°

On its yearly path around the sun, the earth is tilted at an angle. It is this tilt that causes the seasons, not the varying distance of the earth from the sun.

The earth's atmosphere

The earth is enclosed inside a thin blanket of air called its atmosphere. This acts as a protective shield between life on the earth's surface and the hostile environment of outer space.

The gases that make up the atmosphere are held close to our planet by the force of gravity. But where did the air that we breathe come from in the first place? The earth formed about 4.5 billion years ago, and at that time it probably had little or no atmosphere. This developed later as a result of volcanic activity, but some of the gases may have come from bodies such as comets crashing into the earth early in its history.

When the earth was very young, it had many more large, active volcanoes than it does today. These ejected enormous quantities of hot gases, which changed over millions of years into the mixture of gases found in the atmosphere today.

The earth's atmosphere consists mainly of nitrogen and oxygen with traces of other gases. One of the most important minor constituents of the air is water vapor. It is water vapor that produces the clouds that give us rain, hail and snow. The atmosphere also contains very small amounts of the gases carbon dioxide and ozone, but their effects are important as we shall see later.

Right The atmosphere consists of five layers, divided mainly on the basis of how the temperature varies with height. The troposphere (or weather sphere) is nearest to the surface, and the temperature falls with height until the tropopause. Above the tropopause is the stratosphere, where the temperature rises again until the stratopause. The temperature falls again in the mesosphere until the mesopause, which is the coldest part of the atmosphere. Above this the temperature rises once more. Above a height of about 300 mi (500 km) is the exosphere, where the atmosphere merges into interplanetary space.

Left This dramatic view of a hurricane from space shows some of the layers that make up the atmosphere. All the weather systems are confined to the troposphere, which extends to only 10 mi (16 km) above the earth's surface.

Height in miles	Layer	Percentage of atmosphere
60	Thermosphere	0.00001
55		
50		0.0001
45	Mesopause	0.001
40		
35	Mesosphere	0.01
30		0.1
25	Stratopause	
20	Stratosphere	1
15	Ozone maximum	
10	Tropopause	10
sea level	Troposphere	100

Seen from space, the earth is partly hidden by swirling masses of cloud. These are found only in the lowest part of the atmosphere, called the troposphere. This shallow band of air extends upward from ground level to an altitude of about 5 mi (8 km) at the poles and about 10 mi (16 km) at the equator. In this layer are found all the gases essential for life, and all the earth's weather systems take place within it.

Above the troposphere, the atmospheric gases get thinner. The upper atmosphere or stratosphere extends to a height of about 30 mi (50 km). It contains a relatively large amount of the gas ozone. This absorbs much of the sun's harmful ultraviolet radiation that would otherwise be dangerous to animal and plant life. Above the stratosphere lies another layer, the mesosphere, and beyond this the ionosphere, which extends into space.

Temperature and pressure

The temperature of the atmosphere changes with altitude above the earth's surface. For example, if it is a warm sunny day on the ground, and the temperature is 68° F (20° C), it will be about −22° F (−30° C) at a height of 5 mi (8 km). The temperature continues to fall until the top of the troposphere, a level known as the tropopause. Above the tropopause lies the stratosphere, where the temperature remains the same for a while, but then increases steadily until the level of the ozone layer.

The temperature is also considerably less in the polar regions than it is at the equator. This is because at the equator the sun is very nearly overhead, whereas at the poles the sun is always low in the sky and its rays strike the earth at a shallow angle. As a result, they are spread over a wider area and become less intense. They also pass through a greater thickness of atmosphere, which further weakens them.

The gases in the atmosphere press down on the earth's surface causing a pressure called air pressure. Although we are unaware of this pressure, it actually presses down very hard – roughly equivalent to a force of 14.7 pounds on every square inch of your body at sea level.

The coldest temperatures are found near the South Pole.

Death Valley in California is one of the world's hottest places. On July 10, 1913, a temperature of almost 134° F (57°C) was recorded.

However, if we travel up a mountainside or ascend in a balloon, we find that the air pressure gets less the higher we go. It is easy to see why this is so. Air pressure can be compared to a pile of blankets on top of your bed. If you have many blankets (air pressure at sea level) they feel very heavy, but if you have only one blanket (air pressure at the top of a mountain) it feels very light. The layers of air press down on the layers below, and this means that the greatest air pressure is felt at ground level. In the stratosphere the pressure falls to almost zero because there is hardly any air above it.

Air pressure is measured with an instrument called a barometer. Most barometers give a reading in millibars (mb for short). At sea level the reading will be about 1,000 mb on average, with extremes of 870 mb (measured in the center of a tropical storm) and 1,070 mb.

A hot-air balloon rises because the air in the balloon is warmer than the surrounding air. The air in the envelope is kept warm by gas burners. If the air is not kept hot, the balloon will sink back to the ground.

Global wind patterns

The earth's wind patterns are a result of different temperatures and air pressure across the globe. Warm air has a tendency to rise, a process called convection. As warm air rises, surrounding cooler air flows in to replace it. The region of sinking, cooler air presses down more heavily on the ground, creating an area of high pressure.

Near the equator, where the heating effect of the sun's rays is greatest, a band of low pressure is always present. However, at the poles, the sun's rays are weak, and the land is cold and the sea frozen. The air above the polar ice caps is being constantly cooled, and it sinks toward the ground. For this reason, areas of high pressure develop over the North and South Poles.

Below The winds that blow over the earth do not blow in straight lines because they are bent by the spinning earth. Winds are pushed to the right in the northern hemisphere and to the left in the southern hemisphere. This is known as the Coriolis effect.

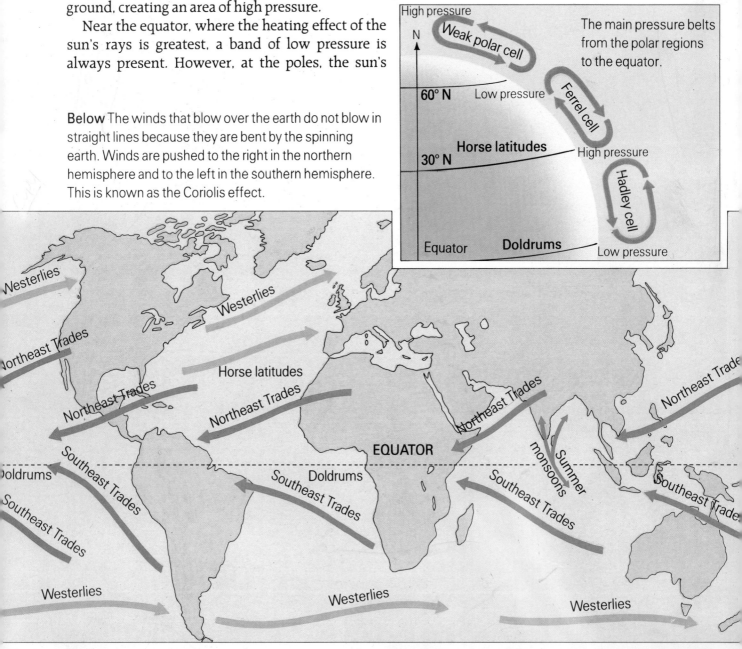

High pressure

N

Weak polar cell

60° N   Low pressure

Horse latitudes

30° N   High pressure

Equator   **Doldrums**   Low pressure

Ferrel cell

Hadley cell

The main pressure belts from the polar regions to the equator.

Westerlies

Westerlies

Northeast Trades

Northeast Trades

Horse latitudes

Northeast Trades

Northeast Trades

Northeast Trades

Northeast Trades

EQUATOR

Summer monsoons

Doldrums

Southeast Trades

Doldrums

Southeast Trades

Southeast Trades

Southeast Trades

Southeast Trades

Southeast Trades

Westerlies

Westerlies

Westerlies

Above and right For hundreds of years, mariners have relied on the wind patterns and ocean currents. In the last century, with the opening up of world trade, clippers used the global winds to sail from Europe and the United States to the Far East.

Above the equatorial regions, the heated, rising air spreads out toward the north and south. At the poles, the sinking masses of air spread out along the lowest regions of the atmosphere. These two movements of air create further bands of high and low pressure in between the equator and the poles. Air at high pressure will always try to move into a neighboring region of low pressure. The winds that blow in every part of the earth are caused by the continual movement of the areas of high and low pressure in the atmosphere.

In the days when sailing ships relied upon the wind to fill their sails, sailors gave names to the regular winds that blew in different parts of the world. The strong winds that blow toward the equator are known as "trade winds." The band of low pressure near the equator where no winds blow was called the "doldrums." Calm air is also found in an area of high pressure across the northern hemisphere, called the "horse latitudes."

Sea breezes and monsoon winds

During daylight hours the land masses and ocean waters are both heated by the sun, but they warm up at different rates. The land warms up fairly quickly, but only to a depth of about two feet. In contrast, the sea heats up rather slowly, because the warm surface layers are continually being mixed with layers of cooler water underneath.

At night, when the land and sea cool down, the reverse occurs. The land cools rather quickly because only a fairly shallow depth of material is storing the heat. The sea, on the other hand, cools slowly – nearly three times slower than the land.

The result of this is that during the day the land is much warmer than the sea, but at night the sea is warmer than the land.

A common feature of coastal regions is the daily, local wind called a sea breeze. This develops on calm, sunny days because the air over the land is warmed faster than the air over the sea. As the air over the land rises, it is replaced by the cooler air from over the sea, producing a sea breeze that blows onto the land. At night, since the land cools faster than the sea, an opposite but weaker land breeze blows off the land.

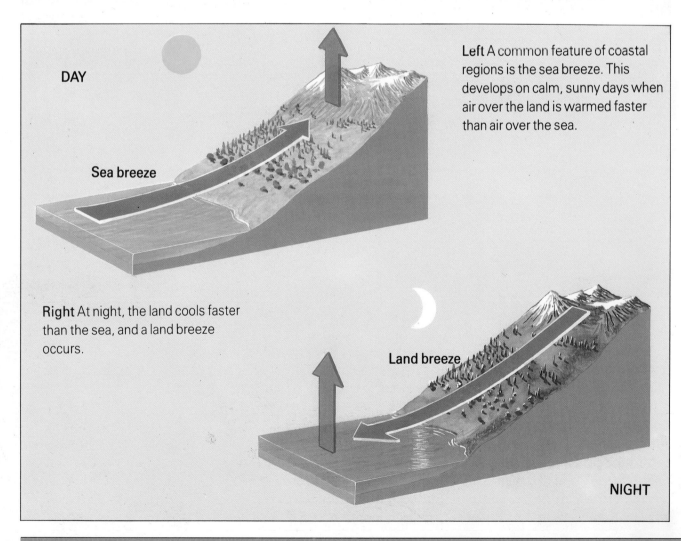

DAY

Sea breeze

Left A common feature of coastal regions is the sea breeze. This develops on calm, sunny days when air over the land is warmed faster than air over the sea.

Right At night, the land cools faster than the sea, and a land breeze occurs.

Land breeze

NIGHT

In some parts of the world, huge land masses heat up so much that seasonal rather than daily reversals of wind direction take place. For example, during the summer months of March to June, the central plains of India are heated intensely. The air above the plains rises, drawing in cooler but very moist air from over the Indian Ocean as a southwesterly airstream. This is the wet summer monsoon. During the winter the opposite happens. The sea is then warmer than the land, and a dry northeasterly monsoon develops.

Monsoon winds are really sea and land breezes but on an immense scale. The arrival of the summer monsoon generates steamy and uncomfortable conditions, violent thunderstorms and torrential rain, essential for the growth of food crops. If the summer monsoon rains are late or interrupted, the crops may fail, leading to widespread famine.

Left In many parts of the tropics, the arrival of the summer monsoon causes torrential rain. This farmer in Indonesia welcomes the monsoon for his rice crops.

Right Mountain valley fog is caused by a katabatic, or downslope, wind. On calm nights, the moist air near the ground cools and becomes heavier. On mountain or hill slopes this air slides downhill to be replaced by air from the same horizontal level alongside. This, in turn, cools and sinks. The cooling process causes water vapor in the moist air to condense and form the water droplets that make up the fog.

The water cycle

About two-thirds of the earth's surface is covered by oceans. The sun's rays heat the oceans, causing some of the water to evaporate into water vapor. This is an invisible gas that mixes with all the other gases that make up the air. Warm air can hold much more water vapor than cold air, and whenever air is warmed over land or sea, it absorbs water vapor. The moist air then rises, but as it does so it cools, and becomes less able to hold the water vapor. Eventually the moist air will be cooled to its dewpoint, the temperature at which condensation occurs. Some of the water vapor then condenses out into very small water droplets that collect together to become visible as clouds. On a cold morning your breath forms a cloud because the warm, moist air exhaled from your lungs is cooled below its dewpoint.

Winds in the atmosphere blow clouds into patches of warmer or cooler air. If a newly formed cloud moves into warmer air, some water droplets evaporate and the cloud gets smaller. However, if the cloud enters a colder region of air, more water vapor condenses into droplets and the cloud increases in size. If the cloud is cooled still further, extra water will condense on the tiny droplets and form raindrops.

Rainclouds forming along coastal mountain slopes on the island of Tahiti in the Pacific Ocean.

Winds blowing across oceans often form clouds when they meet a range of hills or mountains along a coast. Here they are forced upward by air currents, become rapidly cooled, and so raindrops are formed. For this reason the coastal slopes of mountains often experience heavy rain or snowfall. The other side of the mountains is usually dry because the clouds descend to warmer air, the water droplets turn back to vapor, the clouds shrink in size, and the rain stops. This slope of the mountain is often called the rain shadow.

When rain falls over mountains it is collected by streams and runs into rivers and the sea. So, water that once was evaporated from the sea eventually returns there. This is the so-called water cycle, and it is repeated over and over again.

Frost forms when water vapor condenses in freezing temperatures.

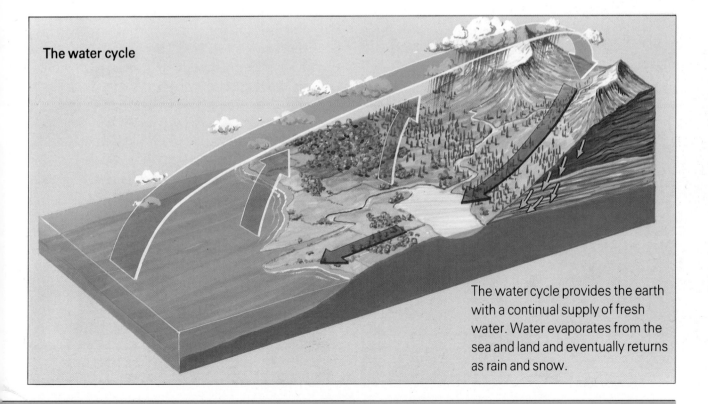

The water cycle

The water cycle provides the earth with a continual supply of fresh water. Water evaporates from the sea and land and eventually returns as rain and snow.

Types of clouds

Clouds appear in a bewildering variety of forms, depending on how and where they arise. The casual onlooker may find it difficult to tell them apart. Most clouds are made up of microscopic water droplets or tiny ice crystals. There are three main types of clouds – cirrus, cumulus and stratus. They were first named by the English chemist Luke Howard in 1802. The three basic cloud types can combine to produce other types, such as stratocumulus, cirrostratus and cirrocumulus. There are about ten varieties.

Lenticular, or wave, clouds sometimes appear in the upper regions of an air stream that has been forced over mountains, when the air is fairly stable.

The familiar cumulus clouds often take on the appearance of absorbent cotton or billow out like enormous cauliflowers. There are three distinct sizes of cumulus clouds: the small or "fair-weather" cumulus, which may heap up in clusters; the towering cumulus congestus; and finally the magnificent cumulonimbus, or thundercloud. Cumulus clouds are composed of tiny water droplets that form as warm, moist air cools and condenses as it rises in the upper atmosphere.

Stratus, or layer, clouds appear as a gray, rather shapeless layer or sheet that extends in all directions across the sky. They are usually only a mile thick, but they can be anything up to 600 mi (960 km) wide. Stratus clouds form when a layer of warm, moist air flows under or over a mass of cold air. Several different kinds of stratus clouds are recognized depending on the altitude at which they occur. Cirrostratus are the highest level clouds, at 3 to 8 mi (5 to 14 km) above ground level, and appear like a transparent, thin whitish veil. Then come the blue-gray altostratus clouds, a medium level layer between 1 and 4 mi (2 and 7 km) and finally stratus, the lowest clouds, rarely lying more than 1,600 ft (497 m) above ground.

Sometimes you can see clouds that appear as thin, curling wisps very high up in the troposphere – actually at altitudes between 3 and 8 mi (5 and 14 km). These are the cirrus clouds, made up of ice crystals, and popularly known as "mares'-tails," after their fibrous, hair-like appearance.

Right Fair-weather cumulus clouds are typical of stable weather conditions in high-pressure regions. Over land, cumulus clouds form and disperse as the land warms and then cools again with the rising and setting of the sun. Sometimes cumulus clouds can become large and the cloud tops billow out like giant cauliflowers. These swelling or towering cumulus clouds are typical of unstable low-pressure weather.

Above The thin, fine white threads of cirrus clouds occur in warm air that is being slowly lifted over a wide area by the approach of a warm front.
Left Dark and gloomy stratus clouds are often associated with light rain or fine snow.

Hurricanes, typhoons and cyclones

In the tropics and subtropics, there are weather systems that are found nowhere else on earth. Of these, the most spectacular are the great destructive storms known as tropical cyclones. These storms are called hurricanes in the Atlantic Ocean, typhoons in the west Pacific, and cyclones in the Indian Ocean. Other weather systems are unique to the tropics. The most common are the large areas of thunderstorms that occur along a band known as the Inter-Tropical Convergence Zone, or ITCZ.

Tropical cyclones begin in the steamy, late-summer heat of the tropics. This warms the ocean surface and the air immediately above it. High surface winds can then cause rapid evaporation of sea water. The water vapor rises to form clouds, transferring massive amounts of heat into the atmosphere. This mixture of heat and water vapor frequently gives rise to violent thunderstorms, which may later develop into a tropical cyclone.

A hurricane in the west Atlantic is usually triggered by an area of low pressure moving westward. This may have started life as a thunderstorm in west Africa. Typhoons, cyclones and a few hurricanes begin in a different way. They start when southern hemisphere trade winds move in a northwesterly direction toward the ITCZ, reach it and push a slight dent in it. In late summer, when the ITCZ is well north of the equator, a group of thunderstorms in the dent may start spinning around. Northern hemisphere trade winds can then carry off this developing storm.

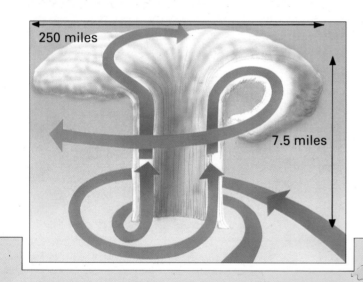

250 miles

7.5 miles

Above A cutaway view of a tropical cyclone. The central region, or eye, is usually about 6 mi (10 km) in diameter. Here, where the pressure is lowest, it is usually calm. Around the eye, in a narrow band 15 to 25 mi (25 to 40 km) wide, the winds blow at their greatest speeds. Outside this band, wind speeds decrease gradually with increasing distance from the storm center.

July

0°

January

The difference between a cluster of thunderstorms and a tropical cyclone is the speed it spins around. Storms that form in the tropics can have their spin increased by the earth's rotation. If this occurs, a group of thunderstorms may develop into a tropical cyclone. A steadily deepening area of low pressure, or tropical depression, forms at the center of the cluster of thunderstorms. Seven out of ten of such depressions develop into hurricanes. The depression becomes a tropical storm when its winds reach gale force, that is 38 mph (62 kph).

A hurricane is usually 620 mi (1,000 km wide), and perhaps 5 mi (8 km) high. As air pressure falls in the storm center, the band of highest wind speeds, circling the "eye" of the hurricane, decreases rapidly to as little as 30 mi (50 km) across. Outside this region wind speeds are low, but within it the winds whirl around at over 100 mph (160 kph).

When a hurricane hits the land it lashes the shoreline with storm force winds in excess of 100 mph (160 kph).

Below and left The diagram shows the flow of air in the vicinity of the ITCZ. Along the ITCZ, cold air masses moving from the north and south meet one another. The air from each hemisphere then rises, forming a band of clouds, rain and thunderstorms as it does so. The map shows the extreme seasonal displacement of the ITCZ, which always extends a few hundred miles north or south of the equator.

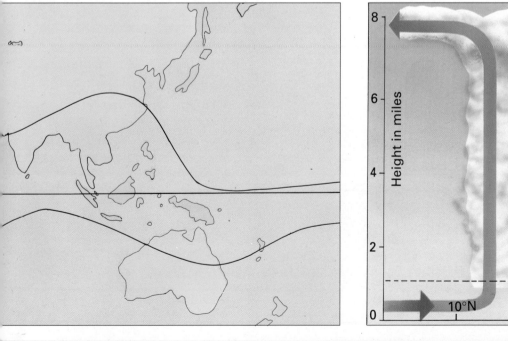

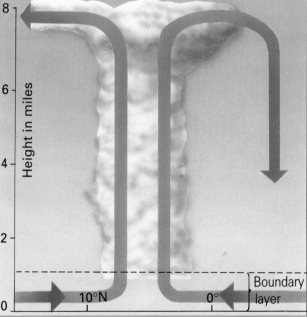

Thunderstorms and lightning

Thunderstorms occur when warm, moist air at ground level starts to rise. As the air rises it cools, and the water vapor within it condenses to form cumulus clouds. In summer the cumulus clouds may develop rapidly upward, producing a cumulonimbus, or thundercloud. These ominous clouds may reach the top of the troposphere where they spread out in the form of an anvil.

Inside a thundercloud, warm air surges upward at speeds that may reach 60 mph (100 kph). The water vapor that the thundercloud carries begins to freeze as the temperature drops to 14° F (−10° C). Water normally freezes at 32° F (0° C), but high in the cloud the water is so pure that droplets can exist at temperatures several degrees lower. These supercooled water droplets collide with tiny particles of dust and ice, and they freeze forming small hailstones. Buffeted by strong winds, the hailstones rise and fall within the cloud, gathering more layers of ice, until they become so heavy that they fall to the ground.

Thunderstorms are nearly always accompanied by lightning and its noisy companion – thunder.

A lightning flash is a giant electrical discharge between the ground and an electrically-charged cloud, or between two clouds. The upper parts of a cumulonimbus cloud usually have a positive charge, while the lower parts contain negative charges. Most lightning flashes last only a fraction of a second. You can often see the lightning flicker. This is because the flash consists of repeated electrical discharges. Each one of these is called a lightning stroke.

A lightning stroke usually begins with a downward leader developing from the cloud. This moves downward along a zigzag path in a series of steps. When the downward leader reaches the ground, a narrow channel (no wider than a pencil) is formed between the charged thundercloud and the ground. A massive electrical current then flows at incredible speed up this channel from the ground to the cloud. This is called the return stroke. It heats the channel to about 54,000° F (30,000° C), causing the surrounding air to expand rapidly. This creates a shock wave like the sonic boom of a supersonic aircraft, and we hear thunder.

Left A spectacular lightning storm showing both cloud-to-cloud and cloud-to-ground lightning strokes. On the island of Java, in Indonesia, there are as many as 300 thunderstorm days each year. In the 25 worst storms, cloud-to-ground lightning strokes occur within about a one-mile radius every thirty seconds for half an hour or more.

Positive charge

Hail

Downdraft

Lightning occurs when supercooled water droplets acquire a negative charge, while other tiny ice splinters or droplets acquire a positive charge. A positive charge collects at the top of the thundercloud and a negative charge at the base. The negatively-charged base of the cloud attracts a positively-charged shadow on the ground, which follows the cloud as it moves along.

Main updraft

Negative charge

Direction of storm

Left Inside a thundercloud, warm, moist air surges upward at high speeds. The air stops rising at a height of between 6 and 8 mi (10 km and 14 km) and the top of the cloud flattens out into an anvil-shaped region of ice crystals.

Tornadoes

In the summer isolated thunderstorms occur from time to time, but sometimes a long line of fifty or more cumulonimbus clouds develops. This is known as a squall line, which can be several hundreds of miles in length. The squall line may move along at speeds of up to 37 mph (60 kph). Squall lines sometimes contain supercell thunderclouds. These are long-lived rotating storms that are particularly violent. The regions of supercell thunderclouds where the wind speed and rotation of the storm are greatest are the spawning ground of nature's most violent phenomena – the funnel-shaped tornado.

A strong, low-pressure region may form within a rotating supercell thunderstorm. This sucks in winds that are rotating around the outside of the storm and causes them to speed up dramatically. Near the center of the supercell, the winds whirl around at incredible speeds that may approach 300 mph (500 kph). Shortly after the high-speed rotation begins, a whirling funnel extends downward from the base of the thundercloud,

sometimes looking like a length of hose. The funnel grows wider and steadily longer until it suddenly reaches the ground in an explosion of flying dirt.

The swirling funnel of the tornado bumps along the ground sweeping up objects like a gigantic vacuum cleaner. The suction is caused by the intense low pressure at the core – it has been known to pluck all the feathers from a chicken! At first grayish-white in color, the funnel grows darker as it sucks up more debris from the ground. Within a large tornado, there may be three or more separate funnels revolving around each other. Sometimes the funnels are like pillars, others are snake-like and twitch repeatedly like the tail of an angry cat.

Severe tornadoes rarely occur outside the United States. Occasionally a tornado traveling over water turns into a waterspout; however, most of these remain quite small. Columns of spinning air, called "dust devils," often develop in desert regions on clear, hot days. Few of these grow large enough to cause damage.

Left Stretching downward from the base of a towering cumulus cloud, the whirling funnel of a waterspout throws spray into the air as it moves across the surface of the ocean. Although waterspouts are not usually as intense as tornadoes, they can produce winds of more than 90 mph (150 kph), although 50 mph (80 kph) is more normal.

Boats have occasionally passed through weak waterspouts with little damage, but larger spouts can completely destroy small craft.

Right The twisting, grayish-white tube of a tornado hits the ground in a flurry of flying dirt, causing its color to change to dark brown. This tornado was photographed in Nebraska, an area where tornadoes are particularly common. The midwestern United States experiences more tornadoes than anywhere else on earth.

Left An aerial view of the damage caused by a violent tornado in Kentucky. Because of their small size and their short lifespans (only a few minutes), tornadoes are extremely difficult to predict.

Blue skies and red sunsets

In the seventeenth century the great scientist Isaac Newton showed that sunlight, also called white light, is a mixture of seven different colors.

Newton allowed a narrow beam of sunlight to pass through a triangular glass prism in a darkened room. The prism bent (or refracted) the beam of sunlight, and stretched it out into a broad band of colored light. Newton noticed seven different colors – we call them the colors of the spectrum. They are red, orange, yellow, green, blue, indigo, and violet. Each color of light travels as a wave, and each has a different wavelength. Red has the longest wavelength, and violet the shortest.

Why is the sky blue? To answer this question we must consider what happens when white sunlight passes through the earth's atmosphere. Here, it is reflected in all directions by millions of tiny particles of dust or microscopic water droplets. This process is known as scattering, and different wavelengths of light are scattered by different amounts. The shortest wavelengths at the blue end

of the spectrum are scattered the most. The scattering of blue light is about ten times as great as that for the longest wavelengths of red light. Other colors are scattered more than the red but less than the blue. Because the sky contains an enormous number of tiny particles and droplets, the scattered light is easily seen, and the sky looks blue. If it were not for this scattering in the atmosphere, a clear sky would look black, except near the sun.

We can also explain why the sun goes a deep red color as it sinks low in the west at sunset. At this time, the sun's rays have to pass through a much greater thickness of the atmosphere than when the sun is high in the sky. As a result, nearly all the blue light is scattered out by the tiny particles in this dense layer. A far greater amount of red than blue light reaches your eyes, and the setting sun appears orange-red in color. Any clouds illuminated by the sun at this time (or just after sunrise) will also appear red or pink, because they are reflecting the red light of the sun.

Left When a narrow beam of light is passed through a triangular glass prism, it is split into a band of colored light, the seven colors of the spectrum.

Right At sunset, the sun's rays pass through a much greater thickness of atmosphere than they do at midday. The colors near the blue end of the spectrum are scattered by tiny particles in the air. As a result, far more red light reaches the eye, and the sun and sky appear red in color.

Rainbows and haloes

Most people have, at one time or another, seen a rainbow. Here the spectrum is produced by the refraction and reflection of sunlight inside raindrops in the sky. This will happen if the sun starts to shine while rain is still falling.

If the sun is fairly low in the sky, and shines on a rain shower (or on fog), raindrops act like tiny glass prisms, refracting the sunlight and splitting it into the colors of the spectrum. Different colors of light are refracted by different amounts as they enter the water in the raindrop. This means that the colors are reflected off the back surface of the raindrop at different angles.

When only one reflection takes place inside each raindrop, a primary rainbow is formed. This has the strongest colors. Since red light is refracted the least and violet the most, the primary bow has red on its outer edge and violet on the inside.

Sometimes two or more reflections occur inside raindrops, forming two or more rainbows. A secondary rainbow, caused by two reflections, will appear above the primary bow in the sky. Its colors are in the opposite order, with the outer edge appearing violet and the inner one red. They are also less brilliant than those of the primary bow because a little light is lost at every reflection.

Mock suns, or sun dogs, are images of the sun formed by the refraction of light through ice crystals.

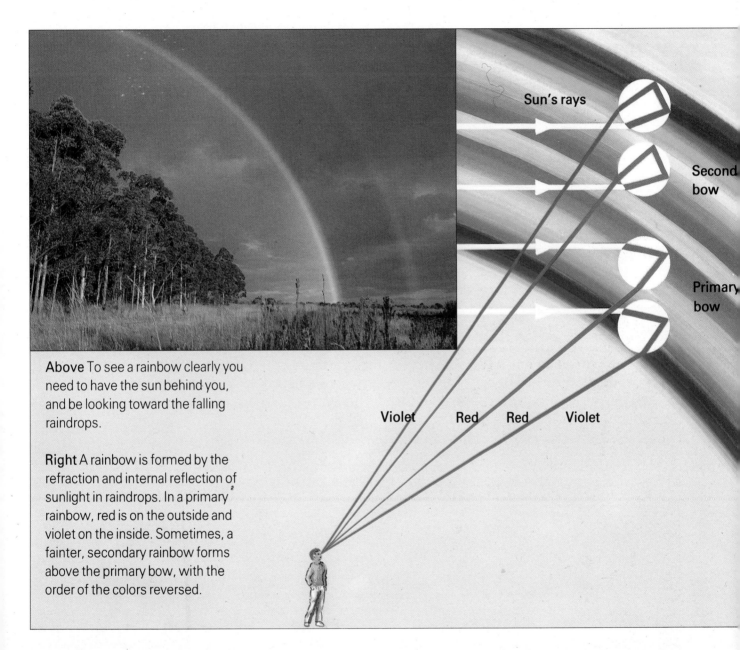

Above To see a rainbow clearly you need to have the sun behind you, and be looking toward the falling raindrops.

Right A rainbow is formed by the refraction and internal reflection of sunlight in raindrops. In a primary rainbow, red is on the outside and violet on the inside. Sometimes, a fainter, secondary rainbow forms above the primary bow, with the order of the colors reversed.

Sun's rays

Second bow

Primary bow

Violet   Red   Red   Violet

From time to time the moon can be seen shining through a thin veil of cloud. If the cloud is composed of water droplets, a series of colored rings can be seen surrounding the moon. This is called a corona, and it is caused by the tiny water droplets splitting the light into colored bands. Ice crystals can also refract light. If the sun or moon shines through a thin layer of ice crystals, such as cirrostratus cloud, a colored ring, or halo, will be seen circling it. The colors are rarely distinct, but this time red is on the inside and violet on the outside. There are two sizes of halo, because different ice crystals bend the light in different ways. The smaller halo is the most common.

Weather systems

As winds blow from high-pressure areas toward low-pressure areas, they cause the pressure centers to move. Their movement from day to day produces the constantly changing weather patterns. Because the air in a high-pressure region (also known as an anticyclone) is slowly sinking, getting warmer and drying out, in summer a high pressure brings warm, dry weather. The reverse is true of a low-pressure region (also called a depression), where the air is rising and cooling, leading to the formation of clouds and rain. Low pressure thus brings wet, stormy weather.

The air is constantly being altered by the surface over which it passes. Air moving over the land will dry out and will warm up or cool down depending on whether the land is warm or cold. Over a warm sea, air will become warm and moist. So at any time there will be large air masses that are either warm or cold. The lines along which these cold and warm air masses meet are called fronts.

A front is the dividing line between the cold, dry air of an anticyclone and the warm, moist air of a depression. Cold air and warm air do not mix together, and when they meet, the heavier cold air forces itself under the warm air, which is pushed upward. This process leads to the formation of clouds along the front, which produce rain. Fronts may travel quite fast, and a period of bright sunshine may be followed quite rapidly by showers. Such changeable weather is common in places like the British Isles, which are affected by fronts sweeping in across the Atlantic Ocean.

Other parts of the world have much more settled and predictable weather. The central parts of large continental land masses are usually covered by regions of low pressure in summer, as the land heated by the sun causes the overlying air to rise. These regions may experience violent thunderstorms in the evening as hot, moist air rises into cooler air, so forming thunderclouds.

A band of cloud marks the boundary between cold and warm air masses as a frontal system advances over Anchorage, Alaska.

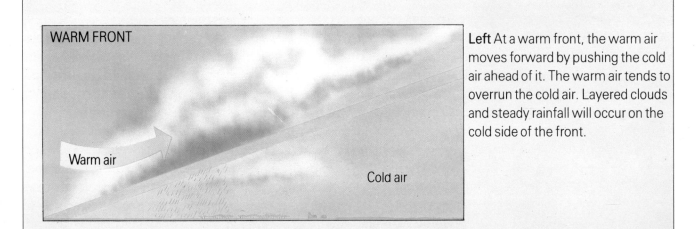

WARM FRONT

Warm air

Cold air

Left At a warm front, the warm air moves forward by pushing the cold air ahead of it. The warm air tends to overrun the cold air. Layered clouds and steady rainfall will occur on the cold side of the front.

Right At a cold front, the heavier cold air moves forward by forcing itself, like a giant wedge, underneath the warm air ahead of it. As the warm air is pushed upward, this process leads to the formation of clouds along the front, which produce rain.

COLD FRONT

Cold air

Warm air

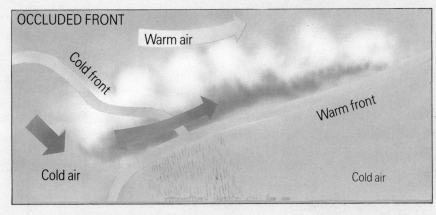

OCCLUDED FRONT

Warm air

Cold front

Warm front

Cold air

Cold air

Left As a cold front overtakes a warm front, the colder, denser air undercuts the warm. The warm air behind the warm front rises up on the cold front ahead of it. Eventually the cold air lifts all the warm air aloft, and the front is then occluded.

Weather maps and forecasts

The people who study the weather and make weather forecasts are called meteorologists. They collect data on weather conditions from weather balloons, weather satellites, ships and local weather stations all around the world. This information is then fed into large computers. The computers make repeated calculations of temperature, air pressure, humidity, wind speed and direction for several levels in the atmosphere, at thousands of separate points over the earth's surface. The results are then plotted onto maps to help in forecasting the weather.

Weather maps show the roughly circular pressure areas, marked "high" and "low." The lines that encircle the pressure centers are called isobars, and they join places with the same atmospheric pressure. As the earth's rotation deflects the wind, northern hemisphere winds blow clockwise around a "high" and counterclockwise around a "low." In the southern hemisphere, the wind directions are reversed. Winds tend to blow in the direction of the isobars, and the closer together the isobars, the stronger the wind.

The fronts, where the regions of high and low pressure meet, are shown by lines with triangles (standing for cold fronts) and semicircles (standing for warm fronts) drawn on them. The fronts move in the direction of the triangles or semicircles. At a cold front, the cold air pushes away the warm air, forcing itself underneath like a wedge. At a warm front, the warm air moves forward pushing the cold air ahead of it. Warm fronts move more slowly than cold fronts because the cold air is denser, heavier and more difficult to push. So cold fronts tend to catch up with warm fronts.

When a cold front meets a warm front, the colder air will undercut the warm. Similarly the warm air behind the warm front will rise up on the cold air in front of it. If the cold air lifts all the warm air aloft, the front is said to be occluded. This is shown on a weather map as a line with triangles on one side and semicircles on the other.

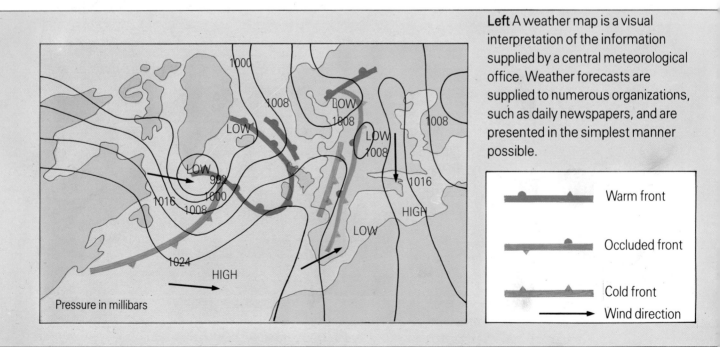

Left A weather map is a visual interpretation of the information supplied by a central meteorological office. Weather forecasts are supplied to numerous organizations, such as daily newspapers, and are presented in the simplest manner possible.

Warm front

Occluded front

Cold front

Wind direction

Pressure in millibars

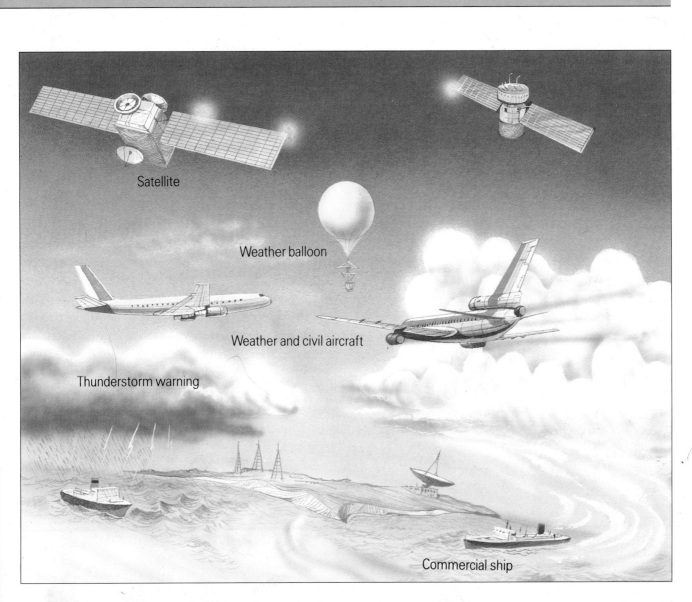

Satellite

Weather balloon

Weather and civil aircraft

Thunderstorm warning

Commercial ship

Satellite images are a great help to local weather forecasters, who usually look no more than 24 or 48 hours ahead. They show cloud patterns over a much larger area than the one for which the forecast is being prepared. An experienced meteorologist checks the images for the position of the various fronts and pressure systems, which are plotted on the weather map. Satellite images are also invaluable for long-range forecasting. By looking at weather patterns on a global scale, useful forecasts for two weeks ahead can be made.

Above When making calculations for a weather forecast, it is necessary to begin with the present state of the atmosphere. There are 5,000 observing stations around the world, which report regularly to data collection centers. More than 1,000 stations send up weather balloons to gather information. Weather satellites are used to determine atmospheric temperature and humidity levels and to monitor cloud type and height. Commercial shipping and aircraft add their reports. All the assembled data are transmitted to weather centers, where the forecasts are made.

Weather satellites

Our knowledge of the weather and the accuracy of weather forecasts has been greatly improved by the use of weather satellites. Orbiting high above the earth's surface, these satellites provide a remote "observation platform" for meteorologists. A satellite image can cover an area of almost 400,000 square miles and is able to show large weather systems in their entirety.

There are several types of weather satellites now in use. They have names like NOAA, GEOS, TIROS, METEOSAT and METEOR, the USSR's satellite series. Low-altitude weather satellites orbit at heights between 500 mi (800 km) and 685 mi (1,100 km). Many of these, such as the TIROS series, have polar orbits. They circle the earth, passing directly over the North and South Poles, while the earth spins beneath them. On each orbit the television camera on board covers a strip about 2,170 mi (3,500 km) wide all around the earth. This strip consists of a sequence of images, taken one after the other, with some overlap. In this way, the earth's surface is scanned twice every 24 hours.

Other weather satellites, such as GEOS, are placed in geostationary orbits. This means that they complete one orbit in exactly 24 hours, keeping pace with the revolving earth, so they appear to "hover" over the same point on the earth's equator. From their much greater distance of 22,195 mi (35,800 km), they can photograph a complete hemisphere of the earth at a time.

Weather satellites take images of the earth by day, and also at night using special cameras that can see infrared or heat radiation. In these cameras, warm areas of land or sea appear dark, and cold areas bright. Warm ocean currents can be identified from the surrounding cooler waters. High cloud that is very cold appears white, and lower levels of cloud appear in various shades of gray, depending on their height and temperature. This allows weather forecasters to distinguish between the different types of clouds. Weather satellites can also measure the density of the atmosphere, the dust content, and even the amount of water vapor, carbon dioxide and ozone.

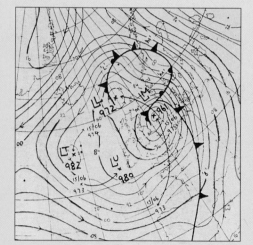

Left A satellite weather picture taken at 0819 on the morning of October 16, 1987, together with the meteorological forecast of the previous day. This shows the center of the "Great October Storm" moving northeastward across the east coast of England toward the North Sea. Gusts of up to 100 mph (160 kph) were recorded at some places.

Below On the TIROS weather satellite there are solar cells surrounding the body of the spacecraft and these provide electricity for the instruments carried. TIROS stands for Television Infrared Orbital Satelite. TIROS weather satellites orbit above each pole of the earth every two hours, circling the earth a dozen times each day. They scan one sixth of the equator each orbit and the entire equator twice each day. The moving satellite takes overlapping images (**right**).

The earth's climate

The word climate is used to describe the average weather of a place over time periods ranging from just a few months to hundreds of millions of years. Two main types of climates are recognized on the earth today. Deep inside the great continental land masses, the temperature range is large and the rainfall is quite small. This is a continental climate, such as you might find in central, eastern Siberia, at the heart of the Asian continent. Nearer the oceans, places experience an oceanic climate. The nearby presence of the sea reduces the range of temperatures, and the rainfall is greatly increased. The many chains of islands in the Pacific Ocean have an oceanic climate.

Sometimes climate can change over relatively short distances. Prevailing winds from the sea rising up over a mountainous island lead to the formation of clouds and rain, such as on the north-eastern coast of Hawaii, where it is very wet. However, as the moist air descends on the other side of the mountains, it is warmed and dried so the clouds and rain clear away. The southwest coast of Hawaii therefore has a comparatively dry climate.

Ocean currents play a major role in controlling the climate of certain places on the earth. You might expect that the closer to the poles you go, the colder the climate becomes. Although generally true, there are exceptions. The British Isles have a

The global system of warm and cold ocean currents plays an important role in controlling the climate of many places worldwide. The temperatures of the ocean surface waters affect the overlying air.

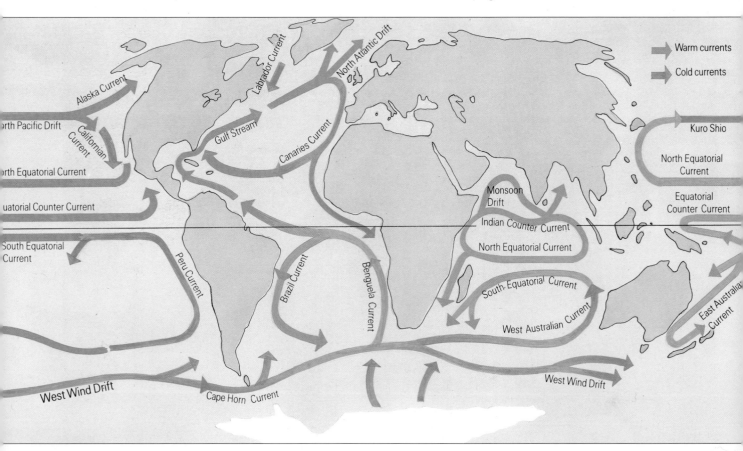

Right The island of Mauritius in the Indian Ocean has an oceanic climate. Oceanic climates are mild and have their extreme temperatures late in the seasons. The temperature range is small, but there is heavy rainfall.
Below Ayers Rock lies deep inside the continental land mass of Australia. Continental climates have distinct seasons and a larger temperature range than oceanic climates. The rainfall is often very low.

very mild climate, while Newfoundland (off the east coast of Canada), which is farther south, has a bitterly cold subarctic climate. The reason is that the shores of Britain are bathed in the warm waters of the Gulf Stream, an ocean current that flows up from the Caribbean, whereas Newfoundland lies within the cold waters of the Labrador Current, which sweeps down from the Arctic Ocean.

These examples show how sensitive the climate of individual regions can be to a range of different factors. Latitude on the earth's surface, the proximity of the ocean, ocean currents, prevailing winds, quantity and type of vegetation, monsoon winds, mountain ranges and the size of land masses all contribute to the variety of climates found on the earth. Even places with the most settled climate, do, on occasions, experience freak weather conditions. For example, on September 1, 1981, snow fell in the Kalahari Desert in southern Africa for the first time in living memory!

Climatic change

At one time it was believed that changes in the earth's climate took place over millions of years. Today, scientists no longer think this is true. There is evidence that even during the past century important climatic changes have taken place. Two distinctly different types of climatic changes are now recognized. Short-term fluctuations are changes that take place over a few decades. Long-term variations are those that take place over thousands or even millions of years.

Scientists have found many ingenious ways of detecting changes that took place long ago. Between 40 and 60 million years ago, important changes took place in the position of the land masses on earth, caused by continental drift. Subsequently, the Antarctic ice grew steadily, reaching its present size about 5 million years ago.

During the past million years, the earth's climate has alternated between warmer and cooler periods. These variations have taken place over periods ranging from 100,000 to 200,000 years apart. In the cold periods, called ice ages, the polar ice sheets and mountain glaciers increase in size, covering considerably more of the earth's surface than they do today. We know that between 120,000 and 75,000 years ago, the earth's climate was fairly warm, a period known as an interglacial. This was followed by an ice age, which led to heavy ice coverage between 65,000 and 25,000 years ago. For the past 20,000 years, the climate has generally been much warmer. In North America, the last great ice sheets disappeared about 10,000 years ago.

The closer we approach the present time, the more evidence we find for climatic change. Between 4000 BC and 2000 BC, world temperatures were about 4 to 5° F (2 to 3°C) higher than now. A cooler period followed from 1500 BC to 500 BC, then a slow warming until AD 800. From 1200 until 1500 the climate was rather unstable, followed by gradual cooling. The period between 1550 and 1850 has been called the Little Ice Age. During that time, glaciers moved farther than at any other time since the previous ice age, and the Arctic pack ice moved farther south. Since 1850, there has been a general warming of the climate, which has continued to the present day.

Three hundred years ago, the climate in northern Europe was much cooler than today. During the cold winters, frost fairs were held in London, England, on the Thames River. This engraving shows a fair held in 1684.

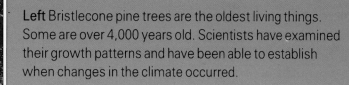

Left Bristlecone pine trees are the oldest living things. Some are over 4,000 years old. Scientists have examined their growth patterns and have been able to establish when changes in the climate occurred.

In the deserts, virtually no rain falls, either because of permanent high pressure or because of protection from prevailing winds by high mountain barriers. In places where the rainfall is already low, such as the Namib Desert seen here, short-term variations in the climate may lead to severe drought and famine.

Causes of climatic change

Climatic change is a complicated process. It may depend on changes in ocean temperatures and currents, the amount of global cloud cover, the extent of the polar ice caps, variations in the sun's energy output, major volcanic eruptions, the earth's orbit around the sun – and also human activities.

Major climatic changes took place about 50 million years ago, when continental drift caused large movement of the land masses. Up to that time ocean currents were prevented from flowing completely around the southern hemisphere. There was probably no ice cap at the South Pole, and the climate was generally warmer everywhere.

Although the southern hemisphere is now rather cooler than the northern hemisphere, this might not always be the case. The land masses in the north form a ring around the Arctic Ocean, separating the polar oceans from the waters farther south. In the future this may lead to a colder climate in the north. This in turn could result in an increase in the amount of northern polar ice. A brilliant white snow or ice surface reflects most incoming sunlight, which leads to further cooling. Even larger ice sheets would be formed, and a new ice age might be started.

Ocean currents can be responsible for both long-term and short-term climatic changes. A warming of the Pacific Ocean, due to a current called El Niño, has been blamed for upsetting the world's weather systems on several occasions.

Continental drift has led to major changes in the earth's climate in the past. About 50 million years ago there were probably no polar ice caps, and the climate was warmer everywhere. Australia and Antarctica were joined together, preventing ocean waters from circulating completely around the southern hemisphere.

The continents 50 million years ago

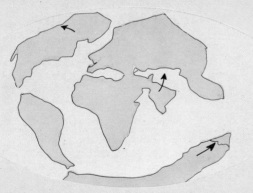

The continents today

In 1980, a short-term variation in weather patterns was caused by the volcanic ash blasted into the atmosphere by the eruption of Mt. St. Helens, Washington.

We tend to regard the sun as a continuous source of energy, even though it has a roughly eleven-year cycle of activity. If the sun remained calm and unenergetic for long periods, important climatic changes would occur. The Little Ice Age of the late seventeenth century may have been caused by a particularly quiet period of solar activity, known as the Maunder Minimum.

Major volcanic eruptions have also been linked with climatic change. If great quantities of volcanic ash and gases are thrown to a height of 30 mi (50 km) in the atmosphere, the dust blocks out the sun's rays, cooling the upper atmosphere. If it is slow to clear, the climatic effects may be severe. Several cooler periods in the earth's past have been linked with major volcanic eruptions.

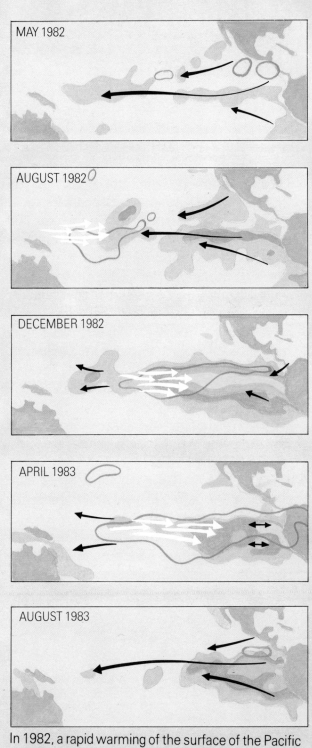

MAY 1982

AUGUST 1982

DECEMBER 1982

APRIL 1983

AUGUST 1983

In 1982, a rapid warming of the surface of the Pacific Ocean, known as El Niño, upset the world's weather patterns. By December, a tongue of warm water stretched for more than 7,450 mi (12,000 km) along the equator. This affected rainfall and wind direction.

The dark blue lines enclose areas of heavy rain. The black arrows show wind directions; the white arrows show the variation from normal.

Man's effect on the climate

Since the gases carbon dioxide and ozone are present in the earth's atmosphere in small quantities, mankind can affect the amount of these gases present very significantly. This may lead to important changes in the earth's climate.

It has been estimated that by the year 2000 the concentration of carbon dioxide in the atmosphere may be 50 percent higher than it was in 1900. Increases in the last thirty years have been mainly due to the burning of fossil fuels (coal and oil), and motor vehicle and factory pollution. Trees and other plants consume carbon dioxide, absorbing it and releasing oxygen into the air. As thousands of acres of tropical rainforest are being irresponsibly destroyed, this may lead to further build-up of the carbon dioxide level.

Carbon dioxide is largely transparent to sunshine, but it traps the heat that rises from the warmed earth. This is called the "greenhouse effect," because the glass in a greenhouse also allows the sunlight in but prevents the accumulated heat from escaping. Most scientists believe that increased amounts of carbon dioxide will warm the atmosphere, but they argue over how this will affect the climate. Some think that a global temperature rise of only a few degrees would cause large-scale melting of the polar ice caps and flooding of low-lying areas worldwide.

The ozone layer shields the ground from the dangerous ultraviolet rays of the sun. Scientists are concerned by the discovery of an enormous "hole" in the ozone layer over Antarctica. This is being intensely studied to find out if it is getting larger. Many people believe that gases called freons are responsible for the damage to the ozone layer. These gases are used in many industrial processes and also in aerosol spray cans as a propellant. When released, the freons find their way into the stratosphere where they react with the ozone and remove it from the atmosphere.

If a major nuclear war occurred, nuclear explosions would send enormous quantities of dust high into the earth's atmosphere. The effects might be so severe that sunlight could be obscured for many months, leading to a significant and protracted fall in temperature. This has been called the "nuclear winter." Major nuclear explosions could also cause damage to the ozone layer. Clearly, we must take great care to conserve the delicate balance that governs the climate of our planet.

When the Antarctic Ozone Hole was first detected in 1986, specially-equipped planes were launched by NASA to collect data. The photo shows a pilot getting into an ER–2 plane loaded with instruments for doing experiments at high altitudes. Scientists hope to find out the extent of the hole and what caused it.

Right Acid rain brings to ground industrial pollution that has been carried by the wind. It has been blamed for causing severe damage to the foliage of pine trees. It can also cause damage to fish stocks.
Below Thick smoke belches from chimneys at a natural gas plant, polluting the atmosphere in Western Australia.

Glossary

Air current A moving flow of air in the atmosphere.

Air pressure The force exerted by layers of the atmosphere on the layers below, and on the ground.

Anticyclone A region of high pressure.

Axis An imaginary line about which the earth spins once every 24 hours.

Barometer An instrument used to measure air pressure.

Condense To turn from a gas or vapor into a liquid by cooling.

Continental drift The slow shift of large land masses caused by movement of sections of the earth's crust.

Convection A circulation of the air caused when warm air rises and is replaced by falling, cooler air.

Corona A series of colored rings formed around the sun or moon as it shines through a thin layer of cloud.

Cyclone The local name for a hurricane originating in the Indian Ocean.

Density A measure of how closely packed together a substance is.

Depression A region of low pressure.

Dewpoint The temperature at which water vapor in the air starts to condense into water droplets.

Downward leader The part of a lightning stroke that develops, in a series of steps, from a charged thundercloud toward the ground.

Drought A prolonged period with hardly any rain.

Evaporation The process by which a liquid becomes a gas due to heating.

Front The boundary between two contrasting air masses.

Glacier A slow-moving mass of ice and snow.

High pressure A mass of descending air around which the wind blows clockwise in the northern hemisphere and counterclockwise in the southern hemisphere.

Humidity The amount of water vapor held in the air.

Hurricane A great storm formed over the west Atlantic Ocean, with wind speeds of more than 75 mph (120 kph).

Ice age A period when the polar ice caps and glaciers advanced to cover a large part of the earth's surface.

Ice sheet A thick layer of ice lying over a large area of land, such as is found in an ice age.

Interglacial A warmer period between ice ages.

Isobar A line on a weather map that connects places having equal air pressure.

Low pressure A mass of ascending air around which the wind blows counterclockwise in the northern hemisphere and clockwise in the southern hemisphere.

Monsoon A wind that changes its direction with the seasons. In southern Asia it blows for six months from the southwest bringing rain and then six months from the northeast.

Occluded front The pushing of warm air aloft as a cold front overtakes a warm front and pushes underneath it.

Ocean current A moving flow of water in the oceans.

Orbit The path of the earth around the sun, or of an artificial satellite around the earth.

Ozone The gas found in the stratosphere that filters out harmful ultraviolet radiation from the sun's rays.

Ozone layer A layer of air containing a relatively high concentration of the gas ozone between 12 and 25 mi (20 and 40 km) above the earth's surface.

Pressure center A region of high or low pressure.

Radiation A wave of energy, such as light, heat or radio energy, sent across space.

Refraction The bending of light when it passes from one transparent medium into another.

Return stroke The massive electrical current that flows between the ground and a thundercloud.

Stratosphere The layer in the atmosphere above the troposphere. It contains the ozone layer.

Supercell A long-lived and particularly violent thunderstorm within which tornadoes may form.

Temperate A moderate, mild climate between the tropics and the poles.

Temperature A measure of how hot or how cold a body is, measured by means of a thermometer.

Tornado A devastating funnel-shaped whirlwind that extends to the ground and destroys objects in its path.

Left Cirrus clouds streak the sky over sea ice at Signy Island in the South Orkneys, near Antarctica.

Tropical cyclone A system of winds blowing inward in the form of a spiral to a center of low pressure, brought about by warm, moist air rising from the ocean.
Tropics A band on the earth's surface stretching across the equator between latitudes 25° N and 25° S.
Tropopause The boundary marking the top of the troposphere.
Troposphere The lowest region of the atmosphere within which all the earth's weather takes place.
Typhoon The local name for a hurricane originating in the western Pacific Ocean (usually the China Sea).
Ultraviolet radiation The radiation found in the rays of the sun beyond the violet part of the visible spectrum.
Vapor The gaseous form of a substance that is normally liquid, for example, water vapor.

Further reading

Ahrens, C. Donald, *Meteorology Today* (West, 1988)

Armbruster, Ann and Taylor, Elizabeth A., *Tornadoes* (Franklin Watts, 1989)

Bailey, Bill, *The Weather* (Macdonald Educational, 1974)

Calder, Nigel, *The Weather Machine* (BBC, 1974)

Ferguson, Mike, and Clark, Colin *Understanding Weather and Climate* (Macmillan Education, 1984)

Hardy, Ralph, et al., *The Weather Book* (Mermaid Books, 1985)

Holford, Ingrid, *The Guinness Book of Weather Facts and Feats* (Guinness Superlatives, 1977)

Lambert, David, *Weather* (Franklin Watts, 1983)

Lambert, David and Hardy, Ralph, *Weather and its Work* (Orbis, 1984)

Lye, Keith, *Weather and Climate* (Macmillan Children's Books, 1983)

Schaefer, Vincent J. and Dax, John A., *A Field Guide to the Atmosphere* (Houghton Mifflin, 1981)

Wachter, Heinz, *Meteorology: Forecasting the Weather* (Collins Publishers, 1973)

Picture acknowledgments

The author and publishers would like to thank the following for allowing their photographs to be reproduced in this book: Alison Anholt-White 5 (top); Bruce Coleman Ltd 11, 15 (bottom), 16, 28, 37 (bottom); Crown Copyright 34 (left); Dundee Meteorological Office 34 (right); Frank Lane Picture Agency Ltd *front cover (main picture)*, 10 (bottom), 17, 19, (all), 22, 23, 24, 25 (bottom), 27, 30, 37 (top), 39 (main picture), 43 (both); GeoScience Features 18; The Hutchison Library 5 (bottom); Mary Evans Picture Library 13 (bottom), 38; John Mason 8, 41; Oxford Scientific Films 29, 44; ZEFA *front cover (inset), back cover,* 4, 5 (main picture), 6 (both), 10 (top), 13 (top), 15 (top), 21, 25 (top), 26, 35, 39 (top), 42. Cover illustrations by Malcolm Walker. All other illustrations by Brian Watson and Tony Gibbons.

Index